MW01703364

Lacing Activities
(Manipulatives That You Can Make)

by
Wanda Maldonado-Strever
illustrated by
Liza Sernett

Publishers
T.S. Denison & Co., Inc.
Minneapolis, Minnesota 55431

Standard Book Number: 513-02004-7
Lacing Activities - Manipulatives That You Can Make
Printed in the United States of America

Minneapolis, Minnesota 55431

Contents

AUTUMN LACING CARDS

Football 5
Leaf 6
Owl 7
Ghost 8
Jack-O-Lantern 9
Turkey 10

WINTER LACING CARDS

Snowman 11
Stocking 12
Christmas Tree 13
Valentine 14

SPRING LACING CARDS

Shamrock 15
Easter Egg 16
Bunny 17
Sun 18
Butterfly 19

OBJECT LACING CARDS

Ball 20
Toy Car 21
Rag Doll 22
Block .23
Space Ship 24
Drum .25
Hot Air Balloon 26

ANIMAL LACING CARDS

Whale 27
Dog 28
Cat 29
Turtle 30
Pig 31
Fish .32
Bear .33
Dinosaur 34

LACING NECKLACES

Halloween Necklace 35
Thanksgiving Necklace 36
Christmas Necklace 37
Valentine's Day Necklace 38
St. Patrick's Day Necklace . . . 39
Dinosaur Necklace 40
Jungle Animal Necklace 41
Farm Animal Necklace 42
Flower Necklace 43
Shape Necklace 44
Alphabet/Name Necklace 45
Number Necklace 48

INTRODUCTION

HOW TO USE THE LACING CARDS

The patterns for the Lacing Cards can be duplicated on tagboard. For younger children the teacher may wish to cut out and punch the holes in the card and then let the child lace the card. Four, five, six and seven year olds will enjoy coloring, cutting out and punching the holes in their own lacing cards.

The children will find the experience of lacing cards much more successful if masking tape is wrapped around one end of a long piece of yarn. The masking tape makes a wonderful "needle" that can't "poke" anyone.

The other end of the yarn can be knotted or can be tied to one of the holes in the lacing card.

Your classroom of children will delight in making holiday and seasonal lacing cards, familiar objects and a terrific variety of animal lacing cards.

HOW TO USE THE LACING NECKLACES

The patterns provided for the Lacing Necklaces will give you a multitude of lacing experiences for your children.

The patterns can be duplicated on tagboard. For younger children the teacher will probably have to cut out the patterns and punch a single hole in the top. Older children will love being able to color, cut out and punch the hole in each of their own patterns.

After the patterns have been prepared and the children are ready to begin lacing their necklaces, provide the children with a multitude of other materials for lacing. Various "do-nut" shaped cereals, macaroni, pieces of cut plastic straws, beads and buttons work beautifully between each of the cut-out patterns the children prepared.

The children will be able to make holiday necklaces, a variety of animal necklaces, a flower necklace for spring or Mother's Day, a shape necklace, alphabet necklaces which can be used for making personal name necklaces and number necklaces. Number counting can be reinforced by counting items as they lace their necklaces.

SPECIAL NOTE

Each of your children will thoroughly enjoy having made their own lacing cards or necklaces to keep and take home. But it is also suggested that the teacher make her/his own set of lacing cards and necklace patterns to keep in the classroom. You may wish to laminate the cards and necklace patterns for extended durability.

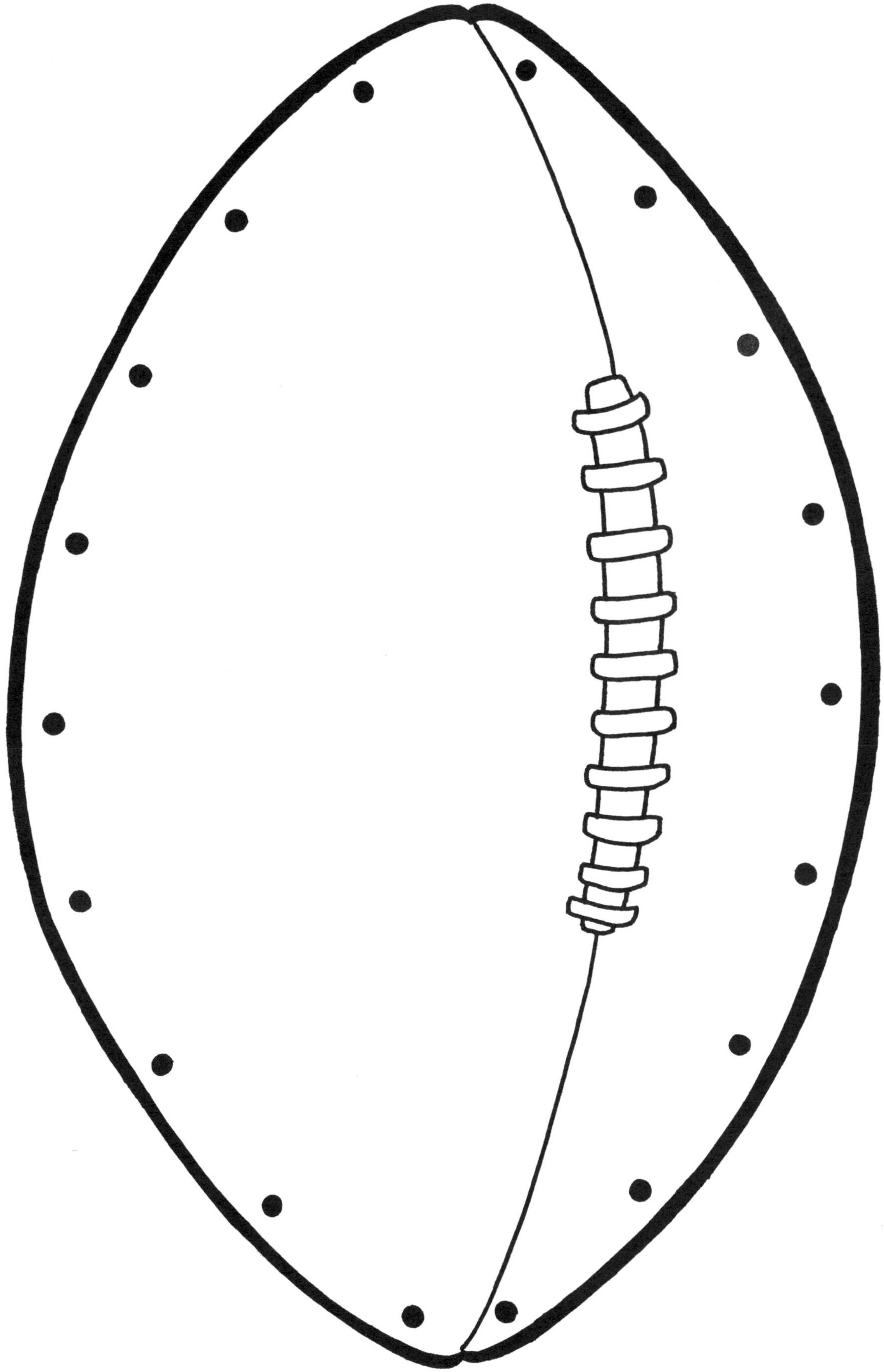

Autumn/Halloween Ghost

Autumn/Halloween Jack-O-Lantern

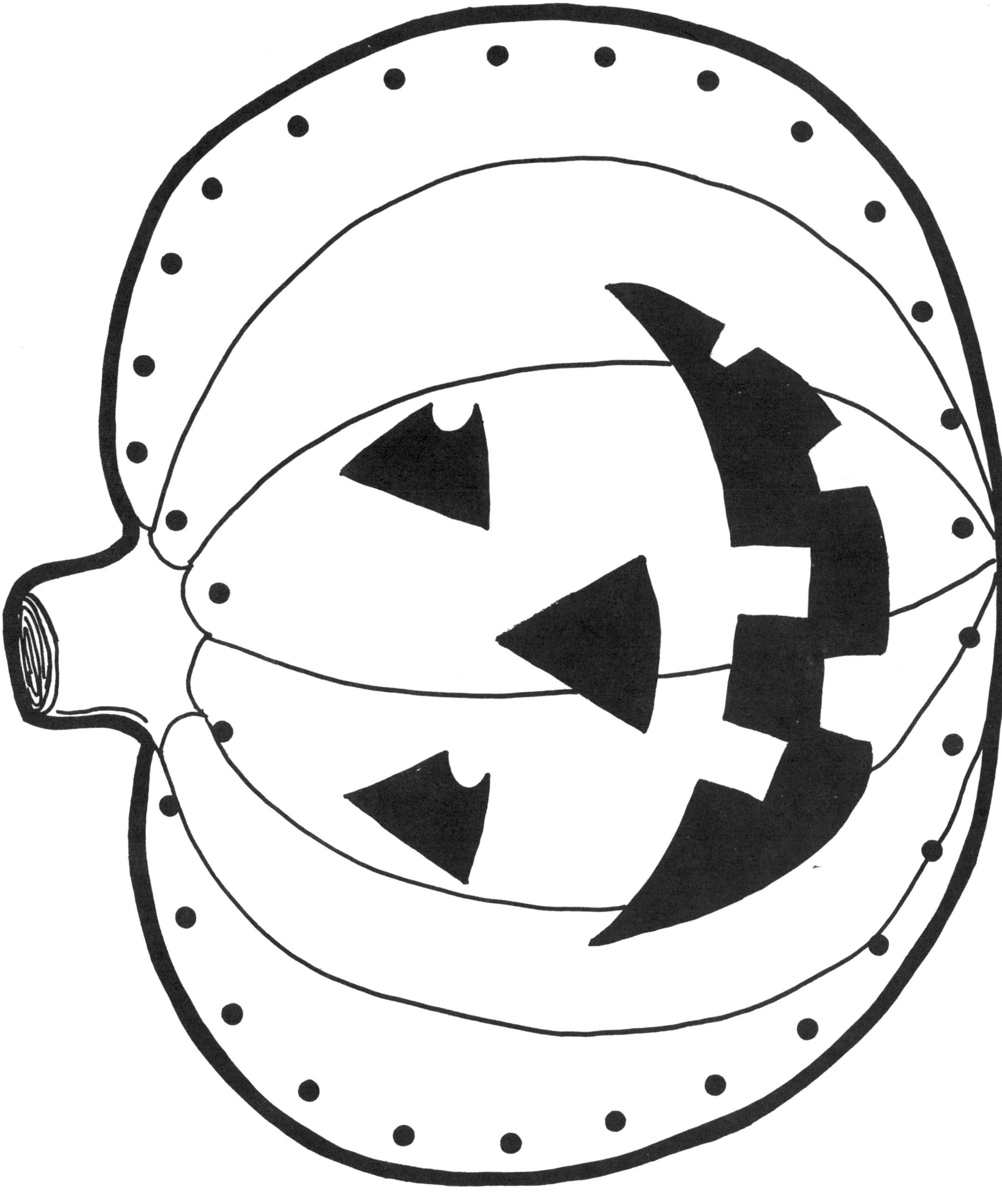

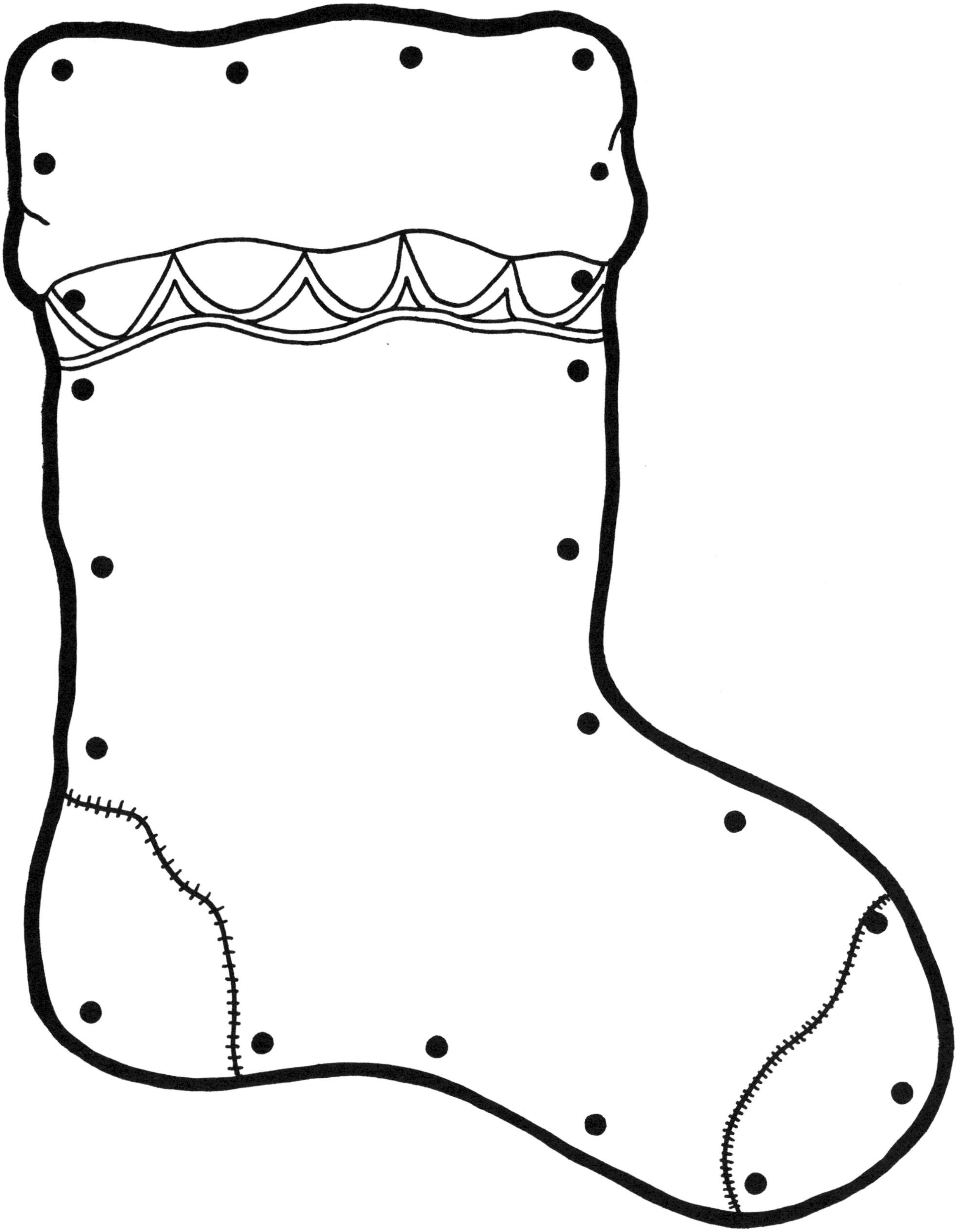

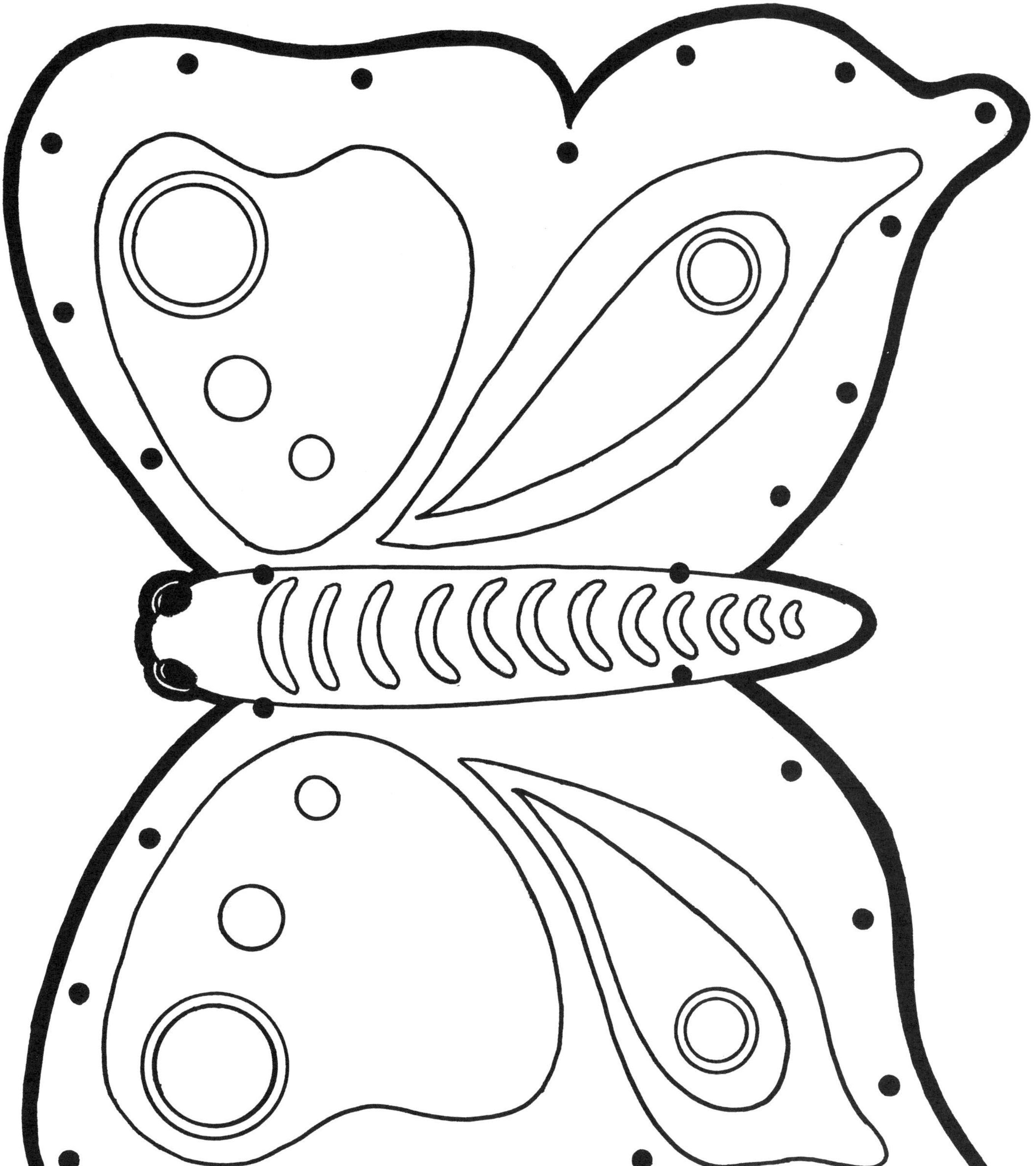

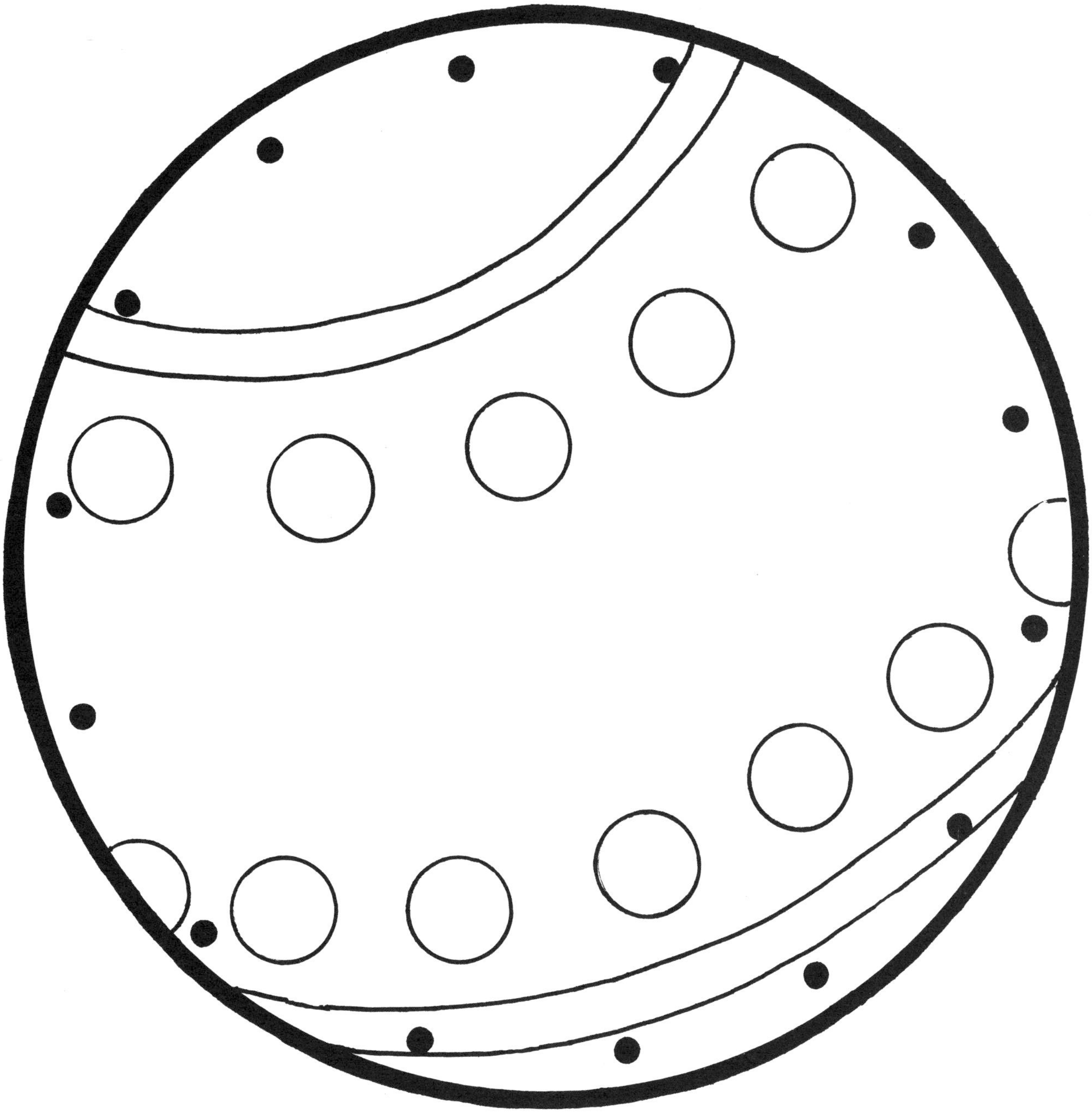

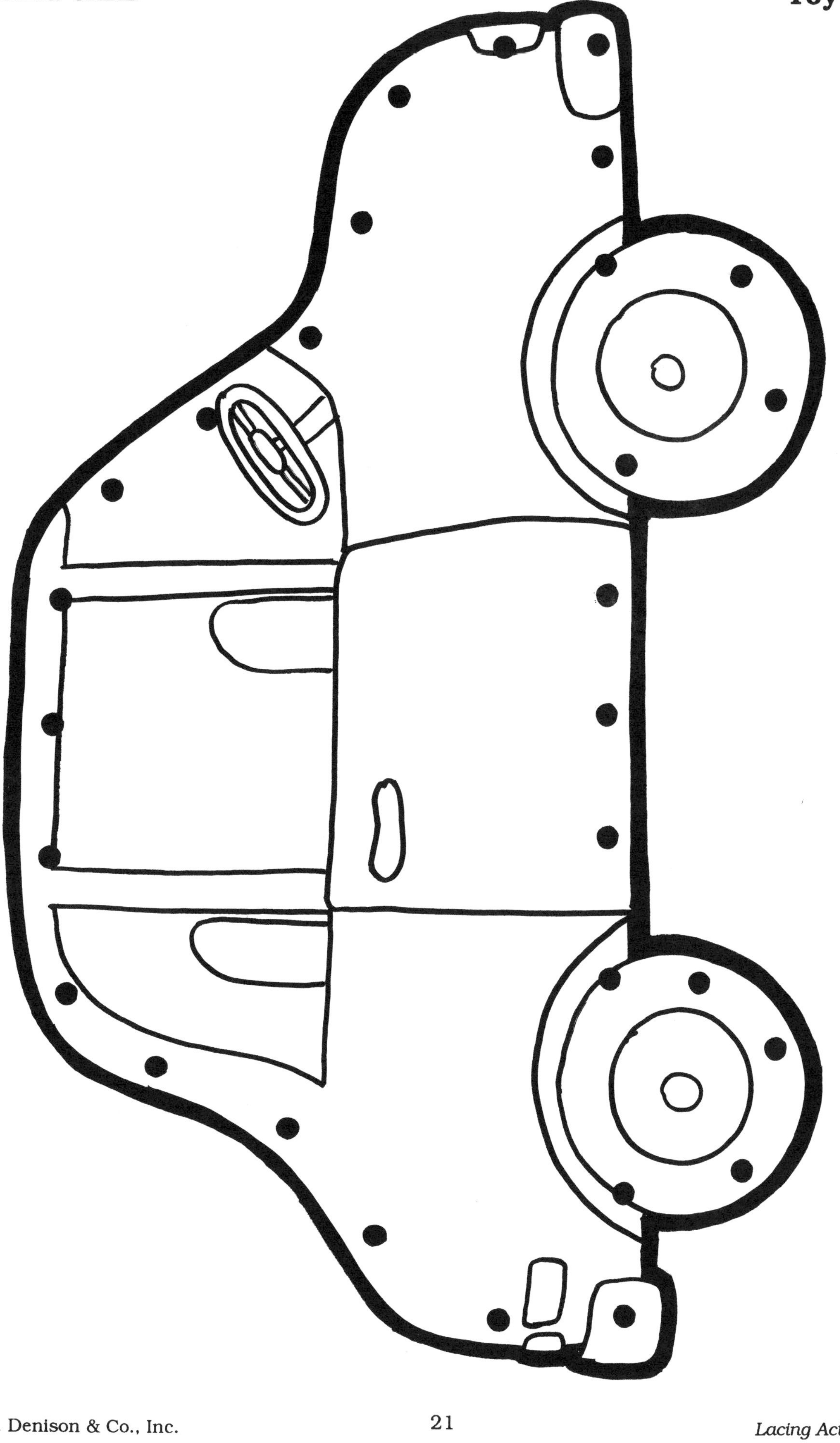

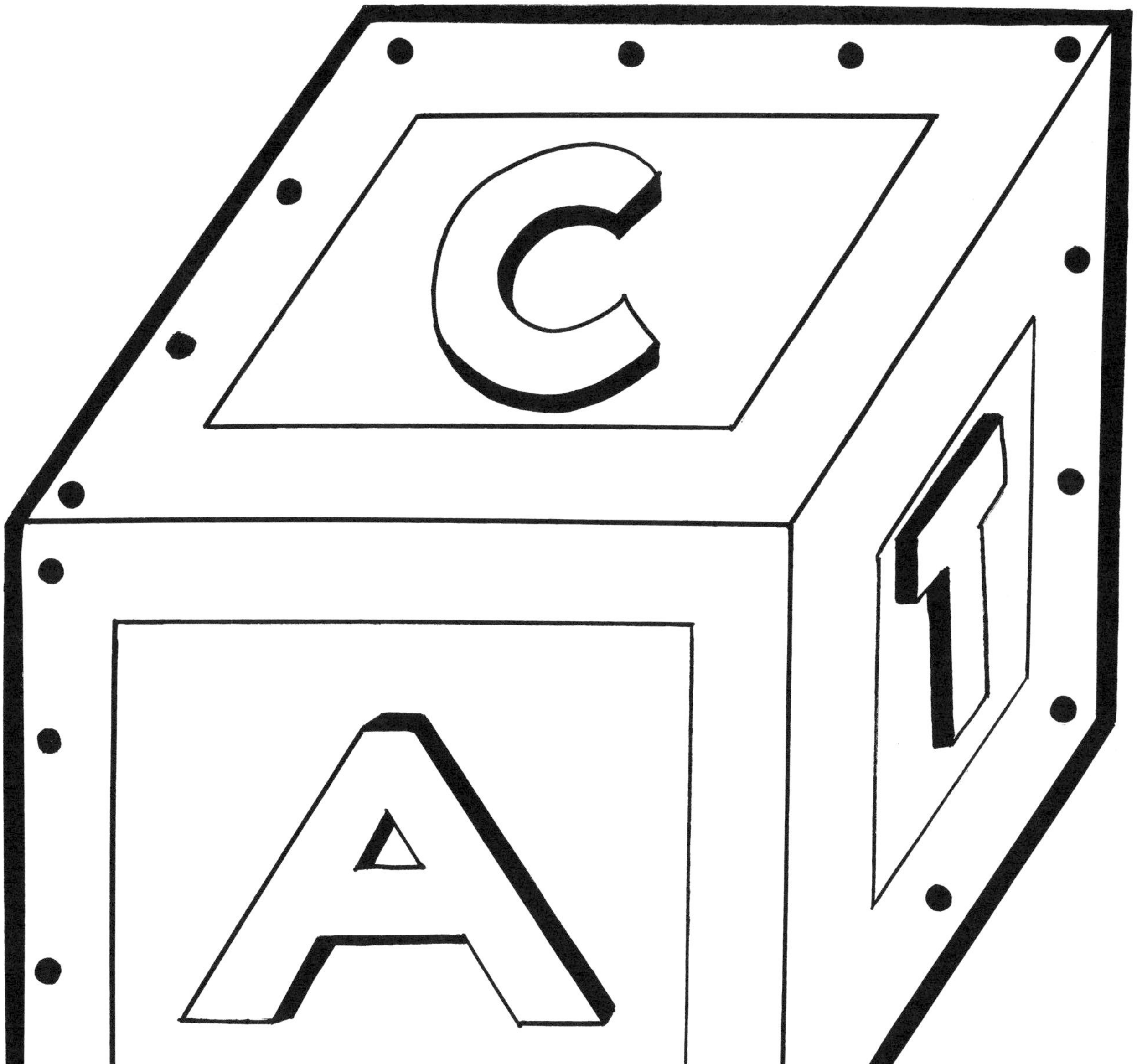
C
A
T

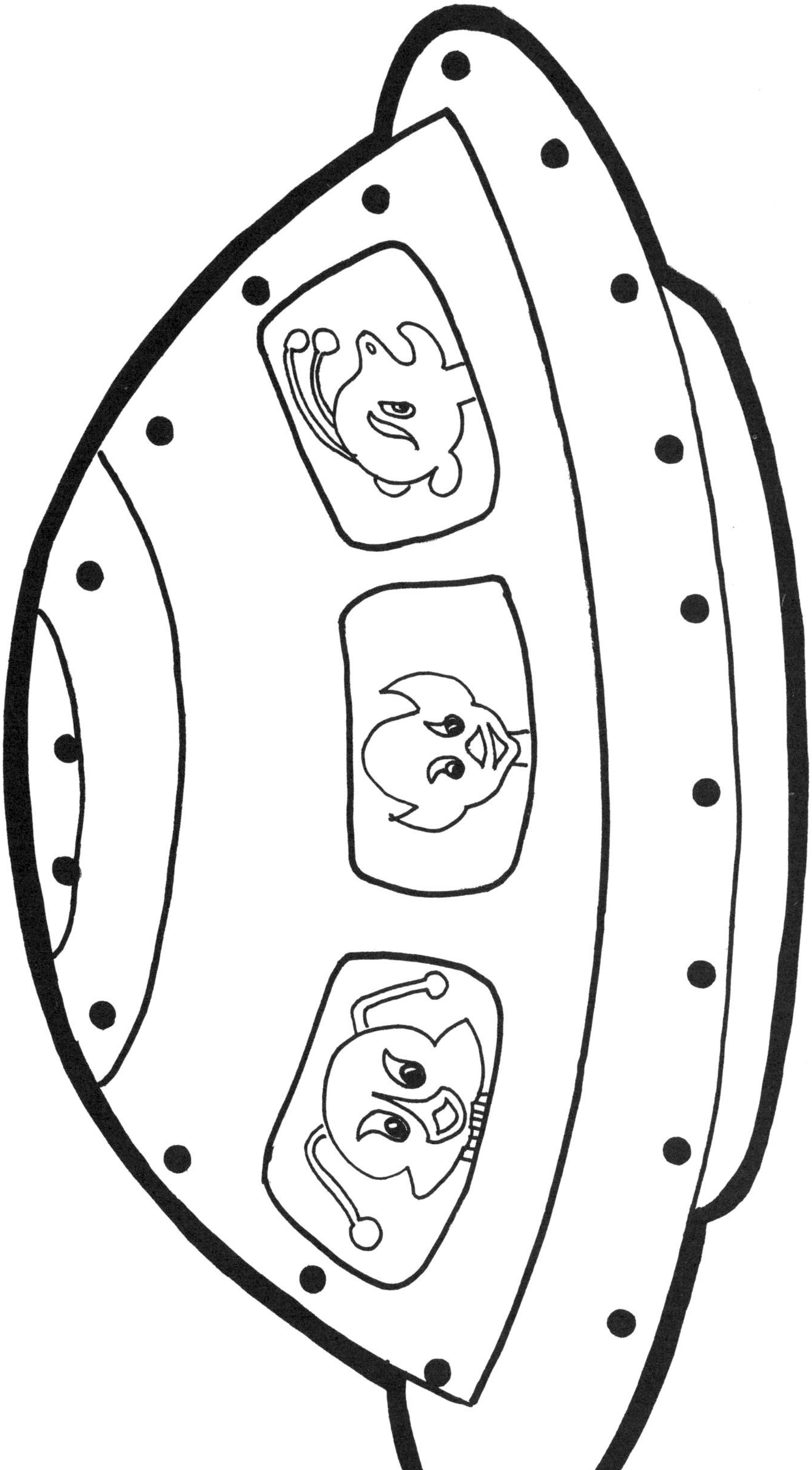

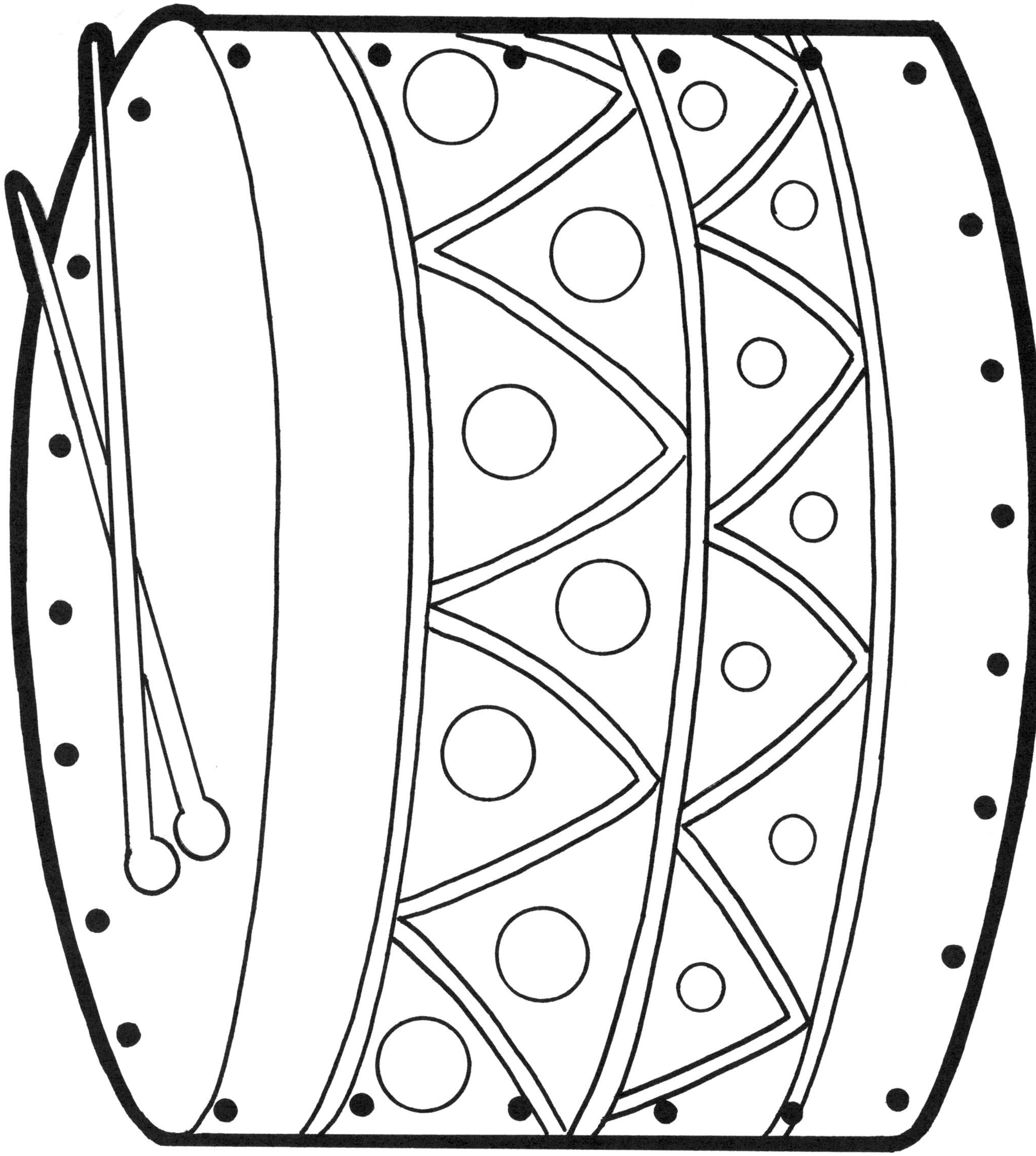

Hot Air Balloon

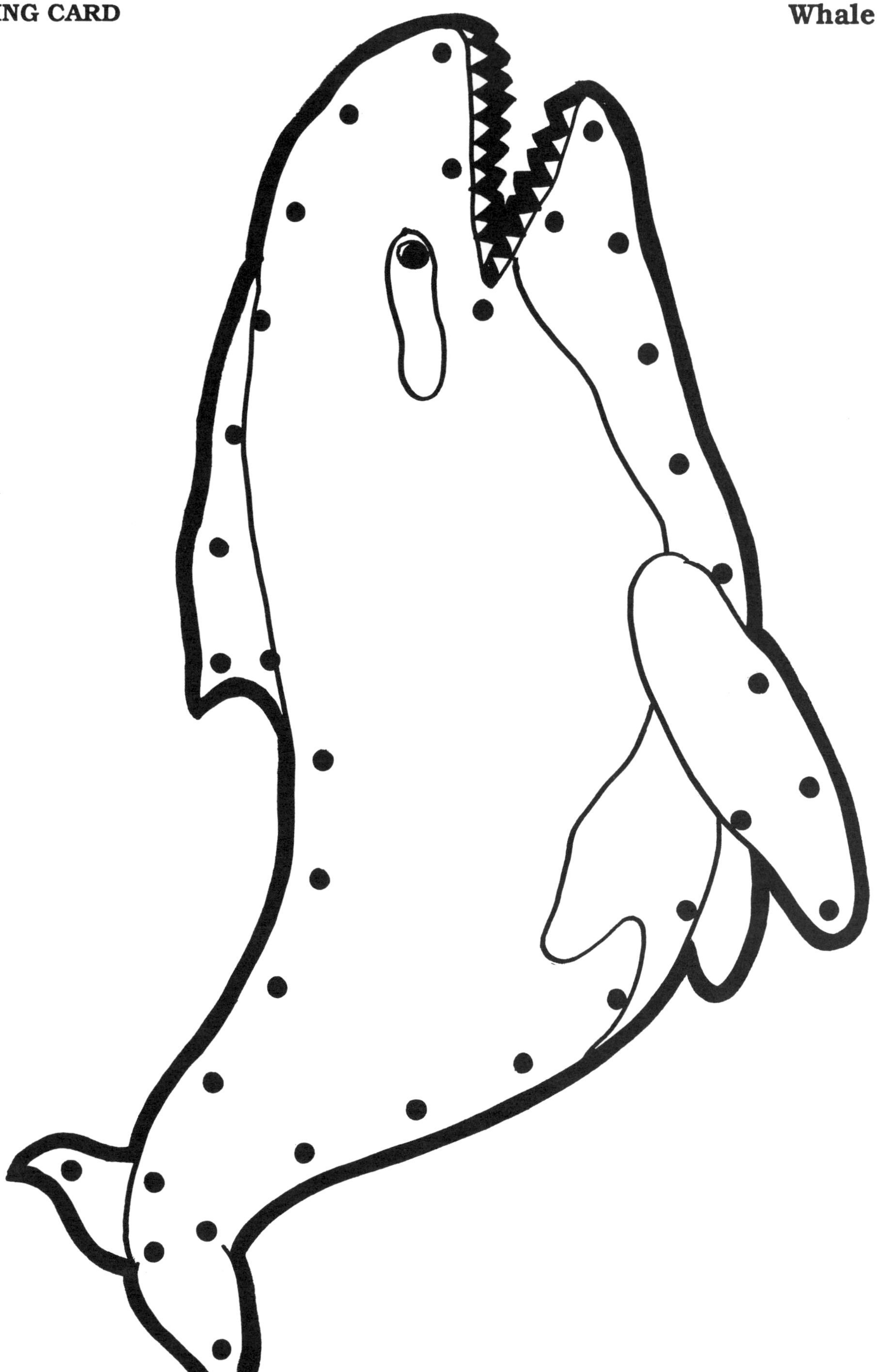

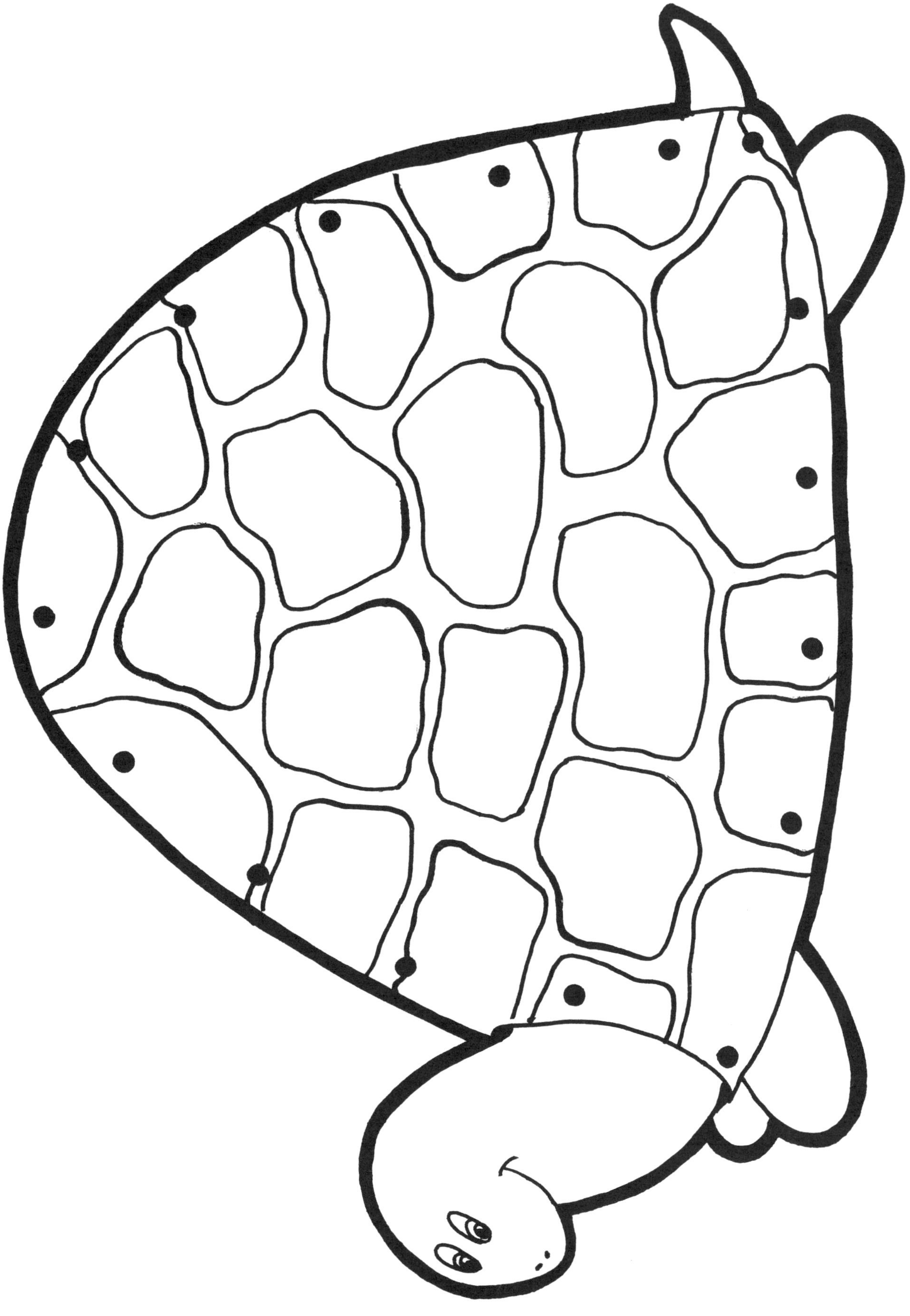

Dinosaur

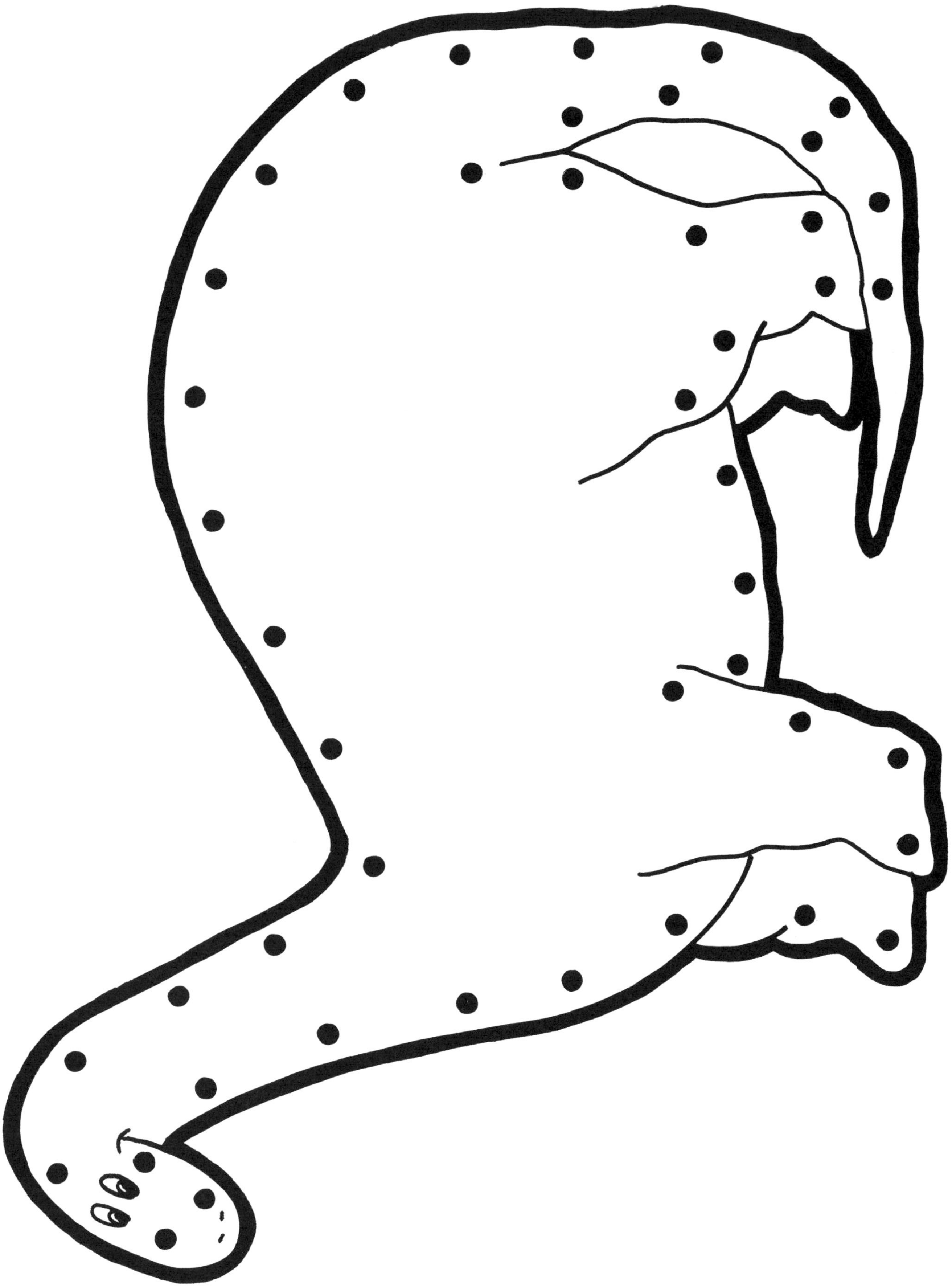

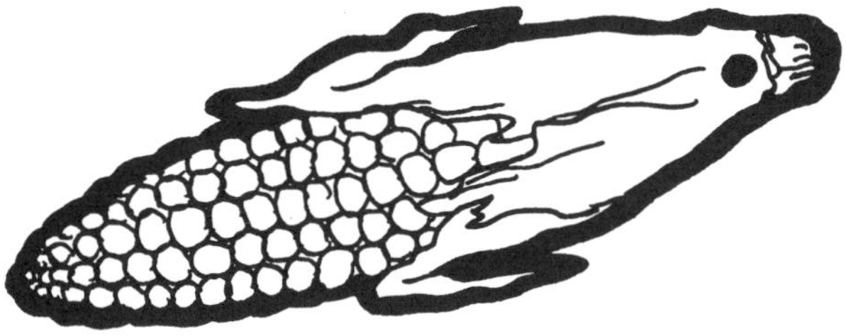

Christmas Necklace

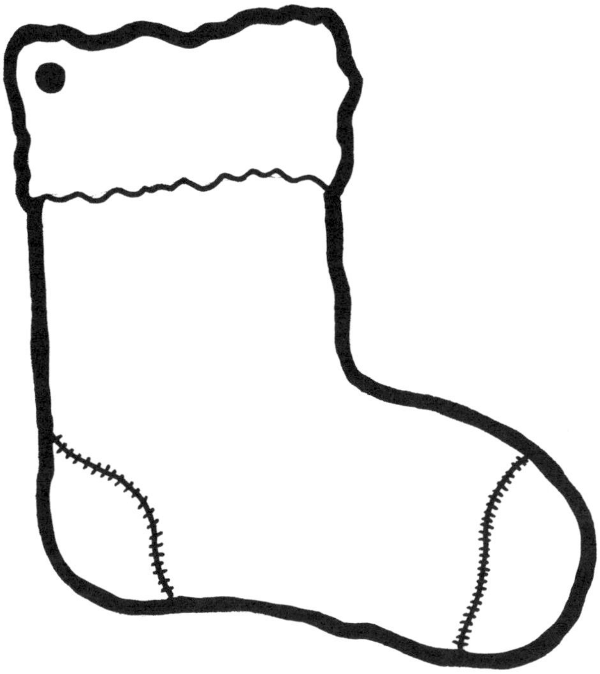

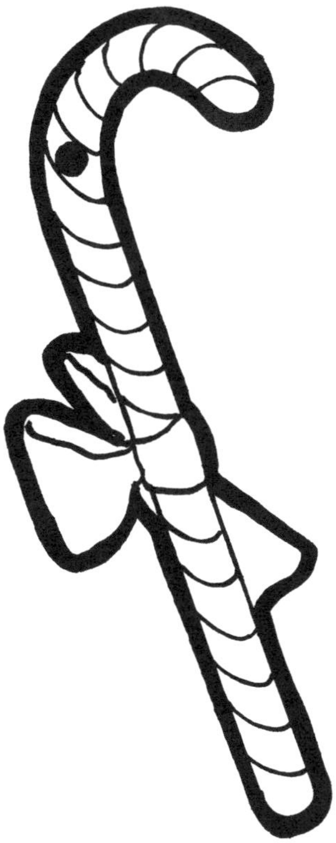

Valentine's Day Necklace

KISS ME
I ♡ U
HUG ME

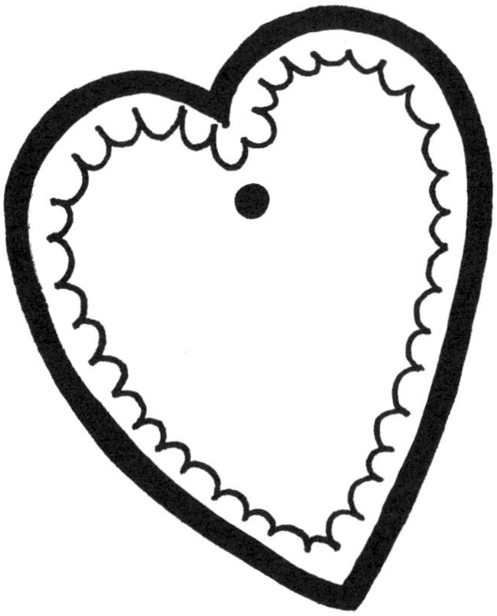

Dinosaur Necklace

Jungle Animal Necklace

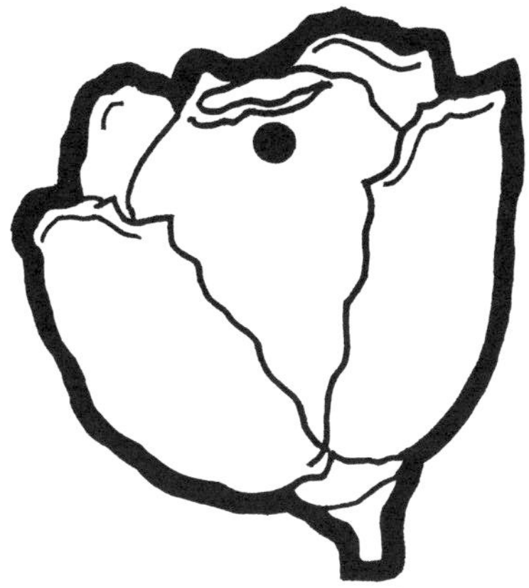

Shape Necklace

Alphabet/Name Necklace

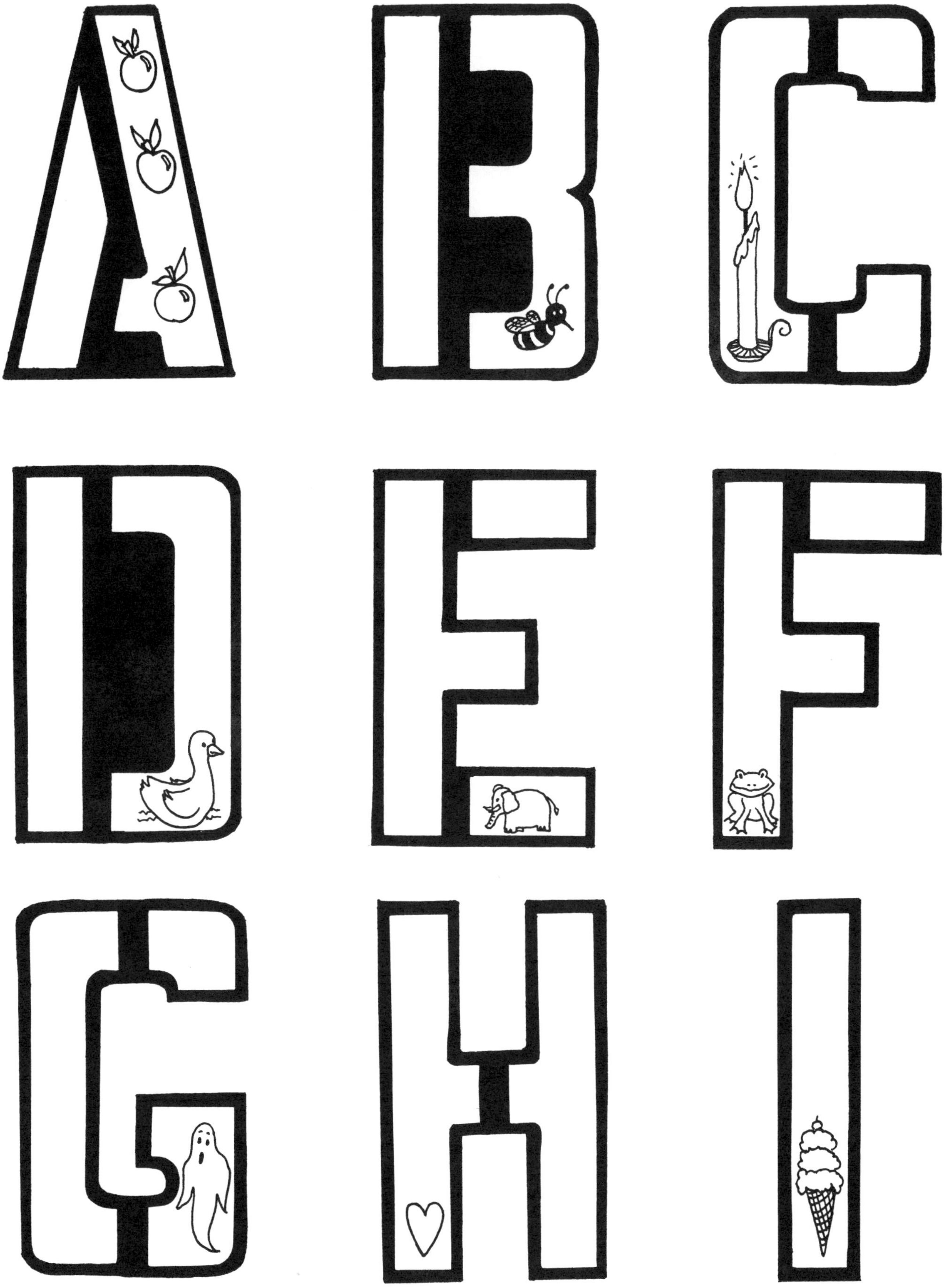

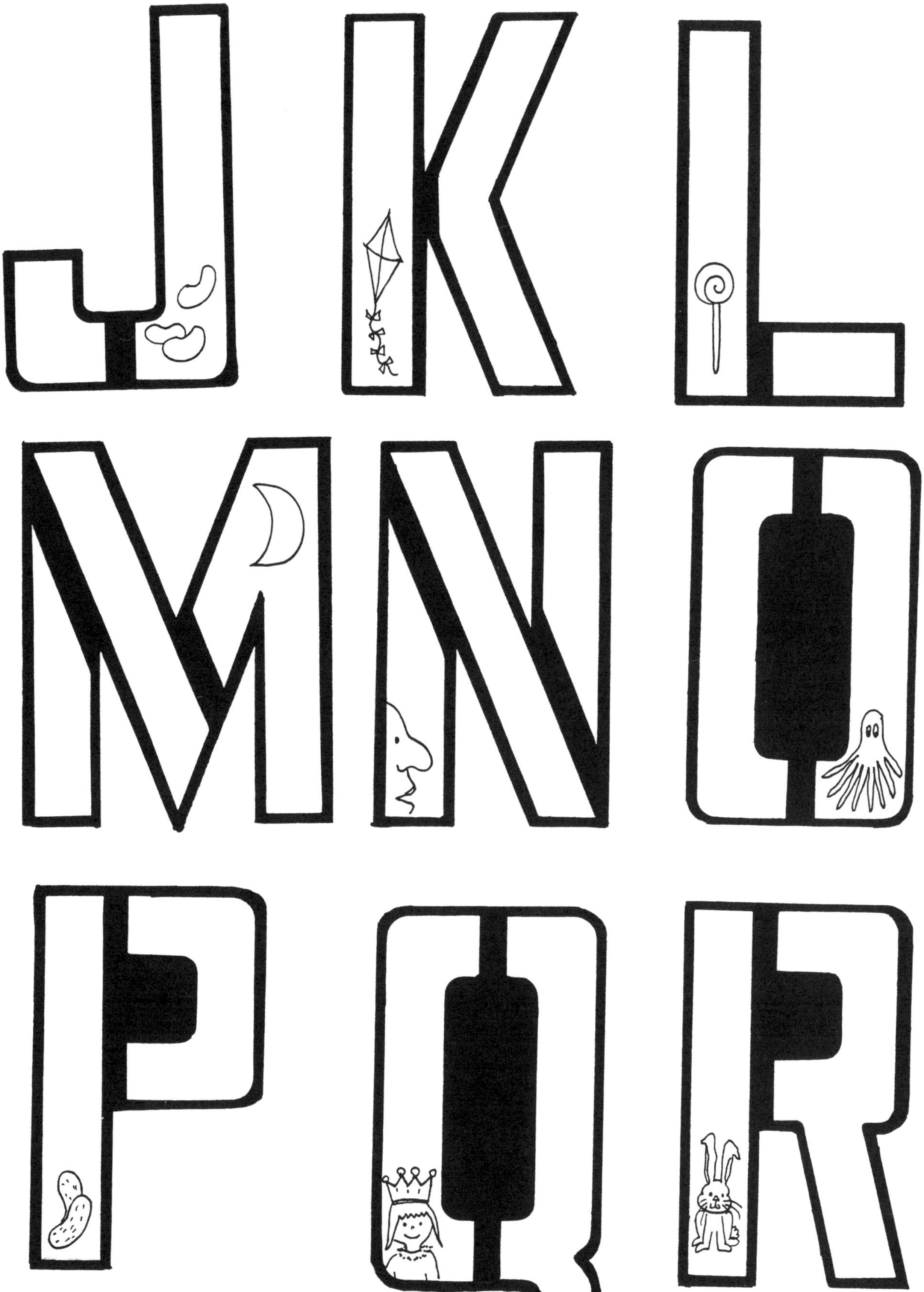

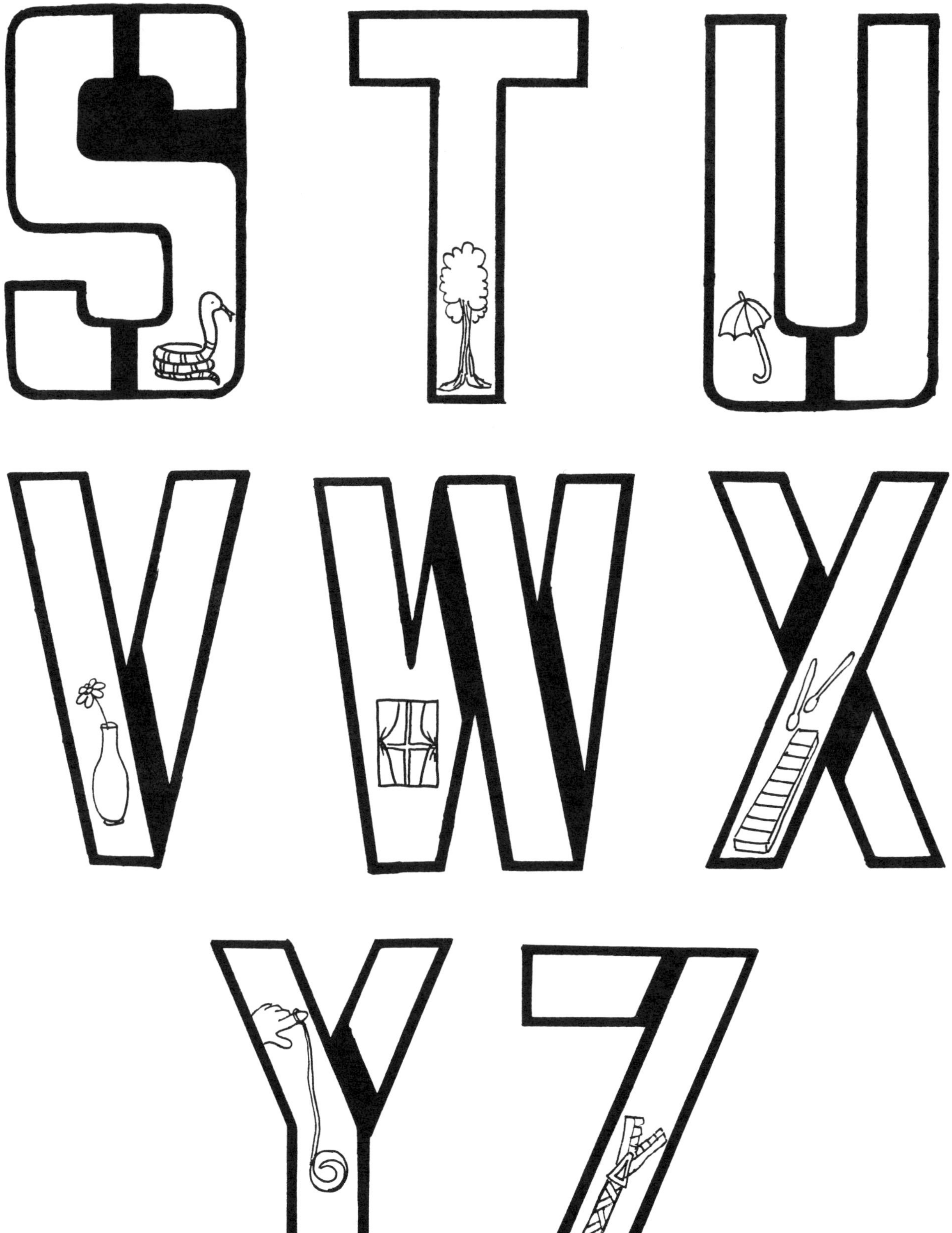
S T U
V W X
Y Z

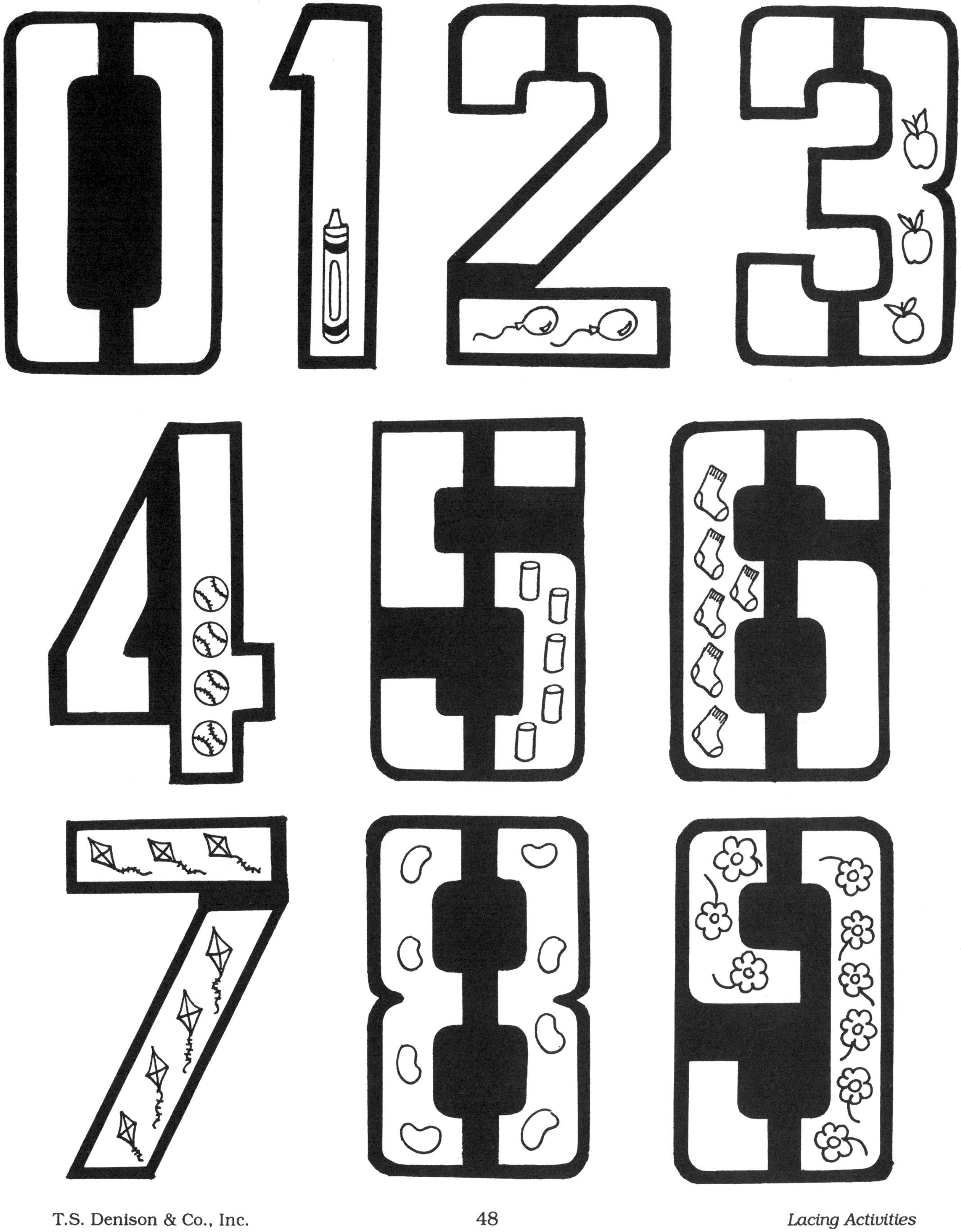